Charles Skelton, Charles Skelton

A brief essay on heat, light, electricity and magnetism

Charles Skelton, Charles Skelton

A brief essay on heat, light, electricity and magnetism

ISBN/EAN: 9783743417434

Manufactured in Europe, USA, Canada, Australia, Japa

Cover: Foto ©berggeist007 / pixelio.de

Manufactured and distributed by brebook publishing software
(www.brebook.com)

Charles Skelton, Charles Skelton

A brief essay on heat, light, electricity and magnetism

A BRIEF ESSAY

ON

HEAT, LIGHT, ELECTRICITY AND MAGNETISM.

BY

CHARLES SKELTON, M. D.

———————◄•►———————

TRENTON, N. J.:

NAAR, DAY & NAAR, "TRUE AMERICAN" OFFICE, PRINTERS.

1875.

CONTENTS.

PREFACE.

I submit this brief essay with much diffidence to the fair criticism of my friends, and of others who feel an interest in the subjects herein discussed. These studies have been pursued for many years in my leisure from more arduous duties, and for the pleasure that I have derived therefrom. Some of the conceptions are new, and all will admit, of the highest importance.

The theories which have heretofore been advanced to explain the laws which govern heat, electricity and magnetism, have not been fully satisfactory to enquiring minds. The theory herein advanced is not presented as fully demonstrated. The full demonstration can only be made by comparing all the known facts with the theory.

The present century has been, emphatically, an age of experiment, and the result is that we have a vast amount of experimental knowledge; but that knowledge lies around us like the material for a magnificent temple, without an intelligent architectural design. We want to know how and where each stone and timber will fit in the great temple of God, the universe.

The vast influence which heat, light, electricity and magnetism exert over all the matter composing the

1

universe, makes this subject all important in the advancement of knowledge. These conceptions, which I have advanced, as far as my knowledge extends, harmonize the facts with the theory.

A full and exhaustive comparison of all our experimental knowledge with this theory would make a large volume :—the attempt here is only to compare the more important and leading facts with this theory ; in this, so far, I see only agreement and harmony, though a fuller investigation of the theory may possibly lead to its rejection.

ESSAY.

CHAPTER I.

ANCIENT AND MODERN THEORIES.

What is heat? This question, to one who has not reflected much on the subject, will appear easy of solution, and unimportant, and yet the ablest minds, from the dawn of science to the present day, have failed to reach a satisfactory solution. The importance of the correct solution equals the difficulties that surround the subject. The solution of this question will lead us into the inner chamber of the temple of science. All the forces of nature are intimately blended with the forces of heat. Light, heat and electricity are so intimately blended as to give rise to the supposition that they are different manifestations of the same force.

Heat and motion are so intimately connected as to give rise to the theory of correlation of forces. This intimate connection has also given rise to what is called the dynamical, or mechanical theory of heat. This theory views heat, as Dr. Tyndall expresses it, as " a mere mode of motion."

Grove, in writing on this subject, says, "Though I "am obliged, in order to be intelligible, to talk of heat "as an entity, and of its conduction, radiation, &c., yet "these expressions are, in fact, inconsistent with the "dynamic theory which regards heat as motion and "nothing else."

Grove very properly says that he "is obliged to speak of heat as an *entity* in order to be understood," for how can a man speak intelligibly of a thing that does not exist? If heat is a nonentity we cannot rea· son about its properties or effects, or even about its convertibility into something else.

The difficulties and importance of the solution of this question are well stated by Tyndall, who says, "The "subject is still an entangled one, and in entering upon "it, we must be prepared to encounter difficulties. In "the whole range of natural science, however, there is "none more worthy of being overcome, none whose "subjugation secures a greater reward to the worker. "For by mastering the laws and relations of heat, we "make clear to our minds the interdependence of natu- "ral forces generally. Let us, then, commence our "labors with heart and hope ; let us familiarize our- "selves with the latest facts and conceptions regarding "this all pervading agent, and seek diligently the links "of law which underlie the facts and give unity to "their most diverse appearances. If we succeed here "we shall satisfy, to an extent unknown before, that

"love of order and beauty, which I am persuaded, is
"implanted in the mind of every person here present.

"From the heights at which we aim, we shall have
"nobler glimpses of the system of Nature than could
"possibly be obtained, if I, while acting as your guide
"in the region which we are now about to enter, were
"to confine myself to its lower levels and already
"trodden roads."

Tyndall's conceptions of heat, as embraced in the
dynamical theory, are accepted by many of our ablest
scientists ; yet this theory to my mind falls short of a
satisfactory solution of the question. Theories have
arisen and fallen at various times from the earliest
periods of history. The mechanical is the last theory
advanced, and is now generally received ; but this, to
my mind, must in time, share the fate of all the theories
that have preceded it. I may be accused of presump-
tion in presenting still another theory of heat. This
too, may pass away as a mere fiction of the imagination,
as all others have. Speculations on this subject, however, are not entirely useless, as the past will fully
demonstrate. Kepler made a great many guesses before
he solved the great laws of the solar system. Nearly
all the theories that have been advanced in regard to
heat have contained some valuable truth; as we shall
find by turning back to many of the ancient conceptions.

We can well understand why fire, at a very early date,

should have been regarded as the chief of the elements, and the motive power of the universe. It had long been worshipped as a symbol of the Deity by the Chaldeans; a worship which probably originated with the Scyths; for Zoroaster, who introduced fire-worship among the Medo-Persic races, is supposed to have been a Scythian. Again, Agni, the god of light and fire, was placed first in the Hindu Trinity. The first scientific idea of heat was that it was one of the elementary forms of matter. This conception was sustained by all that was then known. Heat entered into all the forms of matter, and occupied space; wherever it entered into any form of matter the bulk of such matter was increased. They reasoned correctly from the then known facts, and on this theory they harmonized and explained all that was then known on the subject.

When they worshipped the Sun as the symbol of Deity they were not as unreasonable as many of the present day would suppose. The vitalizing and cheering influences of light and heat were felt on every hand. Thus it was, that in the early ages of the world the inhabitants of the earth were led to worship the Sun as the creator and the giver of life, beauty and happiness. Modern scientific men are much prone to under-value the conceptions and labors of the past ages.

The theory which affirms that the world was composed of four elements : earth, air, fire and water, is the oldest physical conception of which we have any knowledge. It certainly existed before the fifteenth century before Christ. It was adopted in India, Egypt and Greece at a very early date. Several philosophers divided fire into purer and grosser parts. In later times fire would come to signify every thing appertaining to ignition, thus light, whether accompanied by heat or not, flame, the heat inherent in all bodies, incandescent bodies, stars, fiery meteors, lightning, and all visible manifestations of electricities, would be included under the term.

The elemental fire of Herakleitos is the mover of matter, the principle of movement, that which produces perpetual changes around us. Fire was the *anima*, the soul, the vivifying spirit. The mythological side of the belief is seen in the story of Prometheus, who is fabled to have stolen fire from Heaven and therewith vivified mankind. The four-element theory was universally accepted during the middle ages, and was only disproved a century ago, when air was proved to be a mixture of two gases, water a mixture of two gases, fire the result of intense chemical action.

The links that bind together ancient and modern physical thought are strong and enduring; and since they have lasted during the rise and fall of many nations, and during the most profound changes in the mode and tone of thought, it is not unlikely that they will endure as long as the chain itself. We shall see, before we close, that these ancient conceptions were no mere idle dreams. There is a vein of truth, tinctured with many errors, running through all these ancient conceptions.

The idea that matter is composed of atoms, though many discard the conception, is so interwoven with all our modes of reasoning, that we cannot reason intelligibly without them. It is true we have never seen one of these ultimate atoms. They have neither been weighed nor measured, and probably never will be; and yet this conception, which is as old as human thought, still holds possession of the human mind. Dr. Dalton in his profound chemical experiments, and his beautiful and harmonious theory of chemistry, made this conception indispensable to all our reasoning.

The idea of an infinitely rarified and all penetrating matter, was entertained in the earliest ages of philosophy, notably in the Hindu systems; it appears to have been recognised as a fifth element nine hundred years before Christ. Aristotle maintained this conception in his philosophy and supposed it to be always in motion, and to be the moving agency of the other elements. In the present day we find it impossible to explain various phenomena, notably those connected with radiant heat, and the polarization of light, without assuming the existence of some ethereal medium. The wave theory of light as generally entertained, rests entirely on this assumption. Light, heat, electricity and magnetism are so intimately blended that we must consider them as emanating from one elementary form of matter, variously modified in its actions and influences. Our ablest scientists admit the existence of ethereal matter, but deny that heat is an elementary form of matter. As knowledge advances, we find that nature's fundamental laws become simplified. A single great law, as in gravitation, will control and explain a multitude of actions. If we can show that the conceptions of the ancients each contain a part of the truth, and can unite their varied conceptions in one we shall evidently reason in the right direction. The conception of heat as an elementary form of matter, now discarded, contained a part of the truth. The conception of ethereal matter, though enveloped in many errors and mysteries, contained a great truth.

One form of imponderable matter will explain the phenomena of heat, light, and electricity. If we can show that we have imponderable matter filling all space, we shall find this conception sufficient to harmonize and explain all of what are called the imponderable forces. Light is the rapid wave motion of imponderable matter. Heat is this matter in excess in ponderable matter. Electricity and galvanism are this imponderable matter in flowing currents.

This theory shows that all the conceptions of the ancients contained part of the truth. They reasoned well as far as their knowledge extended. Their theory of atoms was purely a mental theory, as the mind could not conceive of a whole thing without parts. Dr. Dalton's announcement of definite proportions in all chemical combinations, which is found to be almost universally true, makes this theory indispensable to all sound chemical reasoning.

The theory of the elementary form of heat, and the theory of a universal ethereal form of matter may be blended in one. This ethereal matter, when in excess in ponderable matter is manifest as heat. Instead of having two kinds of matter we here have but one, heat being only a manifestation of the action of imponderable matter. This theory, herein, in part, harmonizes with the dynamical theory of heat. Motion causes heat, but motion is not in itself heat; nor is this imponderable matter heat. It becomes, or manifests itself as heat when, by being set in motion, it accumulates in excess. Thus, in order to have heat, we must have imponderable matter, and we must have motion to disturb the equilibrium of this matter. In this the dynamical theory is partly true. The purely dynamical theory avers that heat is not material, but simply a mode of motion. In order to have motion, we must have matter to move, and force to move it. Motion implies matter and force, consequently, we cannot have heat without matter; heat must, then, be material. The dynamical theory is right in requiring motion to produce heat, and wrong in denying the materiality of heat. We cannot aver that heat is simply a vibratory motion of the particles of ponderable matter, for this assumption would only lead us one step farther in the dark. The question would then be presented, What caused this matter to move? Has ponderable matter the power of self-motion? We must ever stand on the old assumption that we cannot have motion without force applied.

CHAPTER II.

We know that nearly all the forces acting on the earth come to us from the sun. The revolutions of the earth in its orbit, and the earth's diurnal revolutions, are governed by the sun's forces. Day and night, summer and winter, heat and cold, life and death, all come from the same source of power. All these forces must reach us through a medium. That medium is admitted to be interstellar matter. This matter fills all space, and is infused into all ponderable matter. Heat is produced by motion, but motion does not always produce heat. Motion may produce cold as well as heat. The production of heat depends on the kind of motion and on the kind of matter moved. Motion of expansion produces cold, and motion of contraction produces heat. In other words, when we compel matter to occupy smaller space heat is forced out, and when we expand matter heat is absorbed, and cold results. The more rapid these motions the more intense will be the heat or cold. When the pressure is applied very slowly the heat produced may not become sensible, as radiation, conduction or convection may carry it off as fast as produced.

Many persons suppose that the interior of the earth must be very hot, in consequence of the great pressure of gravitation, The recent dredging in the deep ocean has shown that the deep waters are colder than the surface of the ocean. All ponderable matter, when suddenly compelled to occupy smaller space, gives out imponderable matter, or heat; and sudden expansion absorbs this matter, and produces cold. Combustion is a chemical process, by which the matter burned is compelled, by chemical attraction, to occupy smaller space. The imponderable matter thus rapidly forced out we call heat.

The oxygen is the body from which we get the heat. The purer the oxygen the more rapid the combustion, and the more intense the heat. Hydrogen, the lightest gas known, contains the greatest amount of imponderable matter, and, when condensed by combustion with oxygen, produces the greatest amount of heat known by chemical action. The more expanded ponderable matter is, the more imponderable matter it contains; which, when in excess, becomes heat.

Count Rumford's experiments in boring cannon, in which heat was developed, by friction, led to reflections unfavorable to the theory of caloric. Rumford remarks in regard to these experiments, "By meditat-"ing on the results of these experiments we are

"naturally brought to the great question which has so
"often been the subject of speculation among philoso-
"phers namely : What is heat, is there any such
"thing as an *igneous fluid?* Is there any thing that,
"with propriety, can be called caloric ? We have seen
"that a very considerable quantity of heat may be
"excited by the friction of two metallic surfaces, and
"given off in a constant stream or flux in all directions,
"without interruption or intermission, and without any
"signs of diminution or exhaustion. In reasoning on
"this subject we must not forget that most remarkable
"circumstance, that the source of heat generated by
"friction in these experiments appeared evidently to be
"inexhaustible. It is hardly necessary to add, that
"any thing which any insulated body or system of
"bodies can continue to furnish without limitation
"cannot possibly be a material substance ; and it ap-
"pears to me to be extremely difficult, if not quite
"impossible, to form any distinct idea of any thing
"capable of being excited and communicated in these
"experiments, except it be *motion*." These experiments
and remarks of Rumford led to the abandonment of the
theory of caloric. The unlimited supply of heat was the
main point of difficulty in defending the old theory of
caloric. As there was no probable conception advanced
that would solve the difficulty, the conclusion was that
heat was nothing but motion. This conception might
explain the actions going on in the metals excited, but

Rumford's remark, that "This heat is given off in a " constant stream or flux in all directions," leaves the subject as much in the dark as ever. What constitutes these streams that are given off in all directions? Matter may, possibly, exist without motion ; but motion cannot exist without matter, for in order to have motion some thing must be moved. Matter and motion are ever inseparable. Force without matter to act on, would be a mere abstract conception ; or more properly speaking, we cannot conceive of force without matter to act on. Even matter in motion will not always produce heat. The motion of matter that produces condensation will always produce heat, and motion that produces expansion will produce cold. The motion of a body of matter in empty space, if such a thing could be, would not produce heat. Matter that moves on the earth will always produce heat, as it must move in the air, and in doing so, produces condensation of the air, and consequently produces heat. A cannon ball in passing through the air is heated by condensing the air and receiving the heat thus generated. Tyndall says, " That the ideas of the most well informed philosophers " are as yet uncertain regarding the exact nature of the " motion of heat ; but the great point at present is to " regard it as motion of some kind, leaving its more " precise character to be dealt with in future investiga- " tion." To state that heat is motion makes the subject more incomprehensible than if we say that heat is produced by motion. In the latter case motion is the cause, heat the effect. Every motion must have its cause, and must produce some effect. Cause and effect must ever lie at the foundation of every action.

Pure motion without friction, or expansion, or contraction, will not produce heat or cold. How then does motion produce heat? Motion produces heat only in one way, that is by the condensation of matter, and cold only in one way, by the expansion of matter. With these two propositions, which are in all actions verified, we can trace the varied actions which produce heat, and arrive at clear and certain conclusions. In combustion the motions produce condensation, and consequently produce heat. But what is condensed in the combustion of carbon? Carbon is not; a solid is converted into a gas. The oxygen is the body condensed; the heat is thus derived from the oxygen and not from the carbon. In the combustion of hydrogen with oxygen, the bodies combining are both condensed, and consequently the amount of heat is largely augmented. In the motion of a swiftly flying ball the air is condensed, heat is produced, and the ball and air are both heated. The air is immediately cooled again by its motions of expansion to fill the space just occupied by the ball.

Under this mode of reasoning we can trace the kind of motions which produce heat. When we say that motion is heat we confound cause and effect, and bewilder the mind. When we reason from cause to effect, our logic and facts fit together, and make intelligible conceptions.

Thus far we are consistent and intelligible; but the problem, "What is heat?" is not thus far solved. Why, or how, does the condensation of matter produce heat? The correct answer to this last question will lead us into the inner chamber of nature's laboratory. If we admit that we have imponderable matter filling the interstellar spaces and entering into all the spaces in ponderable matter, we find that we have a key that will unlock the great mystery that has ever hung over this subject. If we cause the atoms of ponderable matter to approach each other, and thus condense this matter, heat is always given out. That ponderable matter which is most expanded contains the largest amount of heat. Hydrogen gas is the most expanded form of ponderable matter known, and when condensed by combining with oxygen, in the formation of water, it gives out more heat than any other known body. If this gas is again restored, by the decomposition of water, the same amount of heat is again absorbed.

All ponderable matter is subject to the same law; additions of heat expand, subtractions of heat contract, the bulk of matter. Thus we see that heat, or ether, is the matter that fills the spaces and expands all ponderable matter. Here we see the reason why and the way in which the condensation of matter produces heat.

Heat, then, if this logic be correct, is imponderable matter in excess. This conception removes the objections which were suggested to Rumford's mind when experimenting on frictional heat. The inexhaustible supply of heat by friction could not be accounted for. The conception of universal imponderable matter filling all space, and filling all spaces in all ponderable matter, makes the supply inexhaustible, and makes consistent the theory of the materiality of heat. The conception that this matter was a separate igneous fluid made all the difficulty. This matter may appear hot or cold as the quantity in ponderable matter is in excess of, or less than, the normal quantity belonging to the particular form of matter.

Heat expands all bodies into which it enters, exactly to the amount absorbed, as in the mercury of the thermometer. Is this thing absorbed matter, or is it a motion of the particles of the mercury? When the thermometer is standing at one hundred degrees we have no evidence that the atoms of mercury are any more agitated than when it stands at forty degrees. We have no evidence to show that rapid motions are going on in the mercury,—no evidence to show that the atoms of the mercury are agitated, or that they are expanding themselves by flying against each other from side to side of the tube which incloses them. The bombarding spoken of by Tyndall and others, as going on in the tube, is a pure fiction. There is something that has entered into the mercury, and expanded it ; this thing that causes expansion occupies space, and must be material.

The advocates of the dynamical theory of heat admit the existence of an ethereal medium, but suppose it to be matter in a very attenuated form. They admit that it enters in between the atoms of all ponderable matter, thus becoming universally diffused through matter and space. As this matter has no gravity, some deny its materiality. That which occupies space, and excludes other matter from occupying the same space, must be material.

Ethereal, or imponderable, matter fills all space, not with atoms far separated from each other, elastic and expansible, but with matter unelastic and full. Elastic compressible matter could not expand ponderable matter with a force that cannot be resisted. The force of heat cannot be resisted any more than the motions of the sun. The force of heat rests entirely on its expansive power. Wherever it enters, the particles of ponderable matter must be separated from each other, to give room to be occupied by the particles of imponderable matter. Pressure will force heat out of ponderable matter, and thus cause contraction ; but when heat enters a body, expansion follows in positively certain ratios. These facts are not consistent with the properties of elasticity. Elastic bodies may be compressed without losing any ponderable matter, but whenever such matter is compressed it loses imponderable matter, or heat, and when expanded again must receive ethereal matter, or heat; and these additions and subtractions are in exact ratios.

Every atom of matter in the universe is absolutely unchangeable in size. Imponderable matter, by entering between the atoms of a mass of ponderable matter, will change its bulk, but such matter will ever remain the same in weight and quantity. Carbonic acid, when compressed into the solid form, loses not one particle of its matter, but loses largely of heat, or imponderable matter, and this heat when suddenly forced out becomes sensible, and may enter other matter again and produce expansion.

The conception of elasticity, or compressibility, is not consistent with the received theory of light. Elastic or compressible ether could not convey light to indefinite distances. Wave motions impressed on our atmosphere can only be conveyed to very limited distances, in consequence of the elastic and compressible properties of our air. Attenuated air will not convey sound. At the height of 15,000 feet two men in a balloon could with difficulty hear each other's voices. Sound is conveyed four times faster through water, and eight times faster through solid matter than through air.

The impulses impressed on ethereal matter by the sun and stars, travel at the rate of 165,000 miles per second, and the distances that these impulses pass over are enormous. The sun's distance, though 92,000,000 of miles, appears small compared with the distances of other bodies which send to us their light. Neptune, the most distant of the known planets, is 2,862,000,000 miles distant from the sun. Many of the comets whose orbits are bound to the sun, traverse much greater distances from the sun. Thus Halley's comet recedes to a distance of 3,200,000,000 miles from the sun; the comet of 1811, 36,000,000,000; and that of 1680, 75,000,000,000 miles from the sum. The period of the last named comet is 8,800 years. Still these figures can scarcely be compared to those which represent the distances of the stars.

It was not till 1840 that the distances of the fixed stars were mathematically ascertained. This discovery is, therefore, of recent date, and we are only now beginning to form an approximate idea of the real distances which separate us from the stars. There are stars whose light cannot reach us in less than 100, 1,000 or 10,000 years!

Dr. Schellen says in his recent work on spectrum analysis: "The light which proceeds from these "stars, is the winged messenger which can bring us in- "formation of their being and nature. Spectrum "analysis has made this light into a ladder on which "the human mind can rise billions and billions of "miles far into unmeasurable space, in order to in-

" vestigate the chemical constitution of the stars and
" study their physical conditions.

" With what acuteness, with what delicacy does
" spectrum analysis accomplish this task ? When the
" balance, the microscope, and every other means of
" research at the command of the physicist and chemist
" utterly fail, one look on the spectroscope is sufficient,
" in most cases, to reveal the presence of a substance.
" If a pound of common salt be divided into 500,000
" equal parts, the weight of one of these portions is
" called a milligramme. The chemist is able, by the
" use of the most delicate scales and the application of
" special skill, to determine the weight of such particle ;
" but in doing so he comes close on the limits of his
" power of detecting by chemical means, the presence
" of sodium, the chief element in common salt. But
" if that small milligramme be subdivided into three
" million parts, we arrive at so minute a particle that all
" power of discerning it fails, and yet even this exces-
" sively small quantity is sufficient to be recognized
" with certainty in a spectroscope."

These impulses made by the sun, and especially by the
stars at these vast distances, could not be transmitted
to us through an elastic medium ; or through very
attenuated matter. Elasticity would very soon destroy
the impact and stop the forward motion. Attenuated
matter could not receive and convey the impressions of
matter in such almost infinitely divided particles.

When light is reflected from the surface of an object it gives back the impression in the most minute particulars, as in the taking of a photograph picture. Widely separated particles of matter could not do this. Nor would elastic matter when reflected be capable of giving back exact impressions.

Ethereal matter is subject to all the laws that govern ponderable matter, except gravitation. When light and heat are impelled against a resisting body, it is reflected in the same way, and at the same angle that a solid, liquid or gaseous body would be deflected. A flowing current of imponderable matter, which constitutes the galvanic current, is subject to the same laws that govern fluid ponderable matter. A galvanic current, when flowing through a large wire, will produce neither light nor heat; but if a small wire is intervened that will not carry the whole current, we then get this matter in excessive quantity, which will give us both light and heat.

A body that is positively electrified will contain this matter in excess and will give off currents of this matter. When two bodies positively electrified are brought near each other, they will repel each other. This results from the two opposite currents meeting and repelling each back on the bodies that gave them off.

Here we see a manifestation of the same mechanical laws that govern ponderable matter. We at the same time see in this example, the cause of attraction and repulsion. Ethereal matter when in excess always flows off in currents, in order to restore equilibrium ; and in doing this affects ponderable matter, giving to it the forces of attraction and repulsion.

The existence of imponderable matter is only known by the effects produced. We cannot confine it ; it will not submit to the ordinary test applied to ponderable matter. Professor Cook, of Harvard University, re- marks, in his " New Chemistry:" "Even in our own time, we still hear of imponderable agents." Do we make a mistake when we call heat, light and electricity agents ? Do we make a mistake when we say that these agents or forces are imponderable ? Can we sup- pose that ponderable matter has the power of self motion ? This would be contrary to all the known laws governing matter. If the sun's forces were with- drawn, and the earth left to its innate forces, all actions would soon cease, and the earth become a lifeless, silent, and frozen mass.

This shows that the forces acting on the earth are not innate, but are derived from the sun, through the medium of imponderable matter.

Matter attracts matter in all its tangible forms. Acids attract and combine with alkalis, but this attrac- tive force may be changed or destroyed by the applica- tion of a current of electricity, or by the application of heat. Some of the properties of imponderable matter are perceived by the senses that take cognizance of ponderable matter. Light, heat and electricity, have their peculiar modes of manifesting themselves to our senses, and we have given them proper names ; but are these properties that we thus name, the result of separate and peculiar forces, or are they manifestations of one force ? Here are, apparently, mystery and confusion. The electric force will produce light and heat, and light and heat will produce the electric current, and attractive force. Can separate forces thus blend and change their powers and manifestations ? Attractive force, it will be recollected, is dependent on, and controlled by all the other imponderable forces. That explanation which will elucidate one of these forces, will apply to all.

The light of the sun comes to us through the medium of an ethereal fluid that pervades the space between us and the sun. The facts which sustain the vibratory theory of light are numerous and consistent, and are, generally, conclusive. If then we have a universally pervading medium, which transmits the impressions of light, the same matter may give us the impressions of heat and electricity.

Nature is simple in her operations, using but few laws or forces to accomplish a multitude of ends. Newton demonstrated the unity and simplicity of the force that governs the solar and stellar worlds. The same force that causes the apple to fall to the ground, sustains ten thousand times ten thousand worlds in their everlasting orbits. Newton only demonstrated the unity of a force. Its cause and connection with other forces were not perceived. That which creates and destroys attractive force, must be the cause of attraction. The same cause that operates on chemical attraction, operates on attraction of gravitation. Heat, light and electricity create and destroy attractive forces. In what way do these forces control chemical attraction ? We must not forget the statement previously made that the agent which causes these forces is ethereal matter in motion. This matter, by the force of its impulse, enters between the atoms of ponderable matter and separates them from each other. The atoms of ponderable and imponderable matter cannot occupy the same space at the same time ; consequently one atom must yield to the other.

In order to explain chemical attraction, let us suppose the atoms of ponderable matter to be formed with angular sides, as we see in all crystals. The sides of these planes coming together, and fitting together, so as to exclude all imponderable matter, may cause adhesion of atoms, in the same way that perfectly fitting plates which exclude the air adhere to each other. Perfectly fitting plates adhere with the force of fifteen pounds to the square inch, this being the pressure of the atmosphere. In the case of chemical attraction, a pound of water, resolved into its constituent gases, would be adequate to raise a weight of 5,314,200 pounds one foot high. This enormous force would represent the pressure of ether on all ponderable matter, and would be the true measure of the force of chemical attraction. This adhesion may be destroyed by forcing this ethereal matter into the joints where these atoms are joined, so that the pressure may be on all sides of the atoms.

What evidence have we that the atoms of ponderable matter have sides and angles? All solid crystals have angles and planes, consequently must be built up of atoms possessing planes. Perfectly round atoms would have pressure on all sides alike, and would be free to move in any direction. If we suppose the atoms of imponderable matter to be perfectly round, we can account for the absence of the force of gravity in this matter, as the equal pressure on all sides of the atoms would leave them free to move in all directions. The angular atoms of ponderable matter would be driven to a common centre.

The cause of chemical attraction, as well as the cause of attraction of gravitation, has ever eluded the wisest heads. The conception of currents of force is entertained by many able scientists. The cause of these currents, and their mode of action, are yet unexplained, and the subject still remains as much in the dark as before the advancement of this conception. The mind ardently seeks for enlightenment on this profoundly important subject.

M. Emile Saigey, in writing on this subject, says of the ether: "Thus, this fluid produces attraction in "matter without itself being subject to it; it confers "gravity upon bodies, and itself is imponderable." The way in which this matter confers these properties, is not made clear by this author. If this universal matter has pressing force, like the air which surrounds the earth, we may form some conception of its mode of action.

This theory as to the mode by which the attractive forces are exerted is not advanced as fully free from objection, but as a conception that may lead to fuller investigation of this all-important subject. We should not rest satisfied with our present attainments, but should pursue our investigations with untiring zeal. This conception, here advanced, is worthy of being fully tested.

Lavoisier, in a memoir which he read to the Academy of Paris, in 1775, announced that calcination and combustion are the results of the union of a highly respirable gas (oxygen) with combustible bodies; and, soon after, he proposed the theory that the heat produced during combustion was disengaged from the oxygen. "These two propositions," says Cuvier, "belong to "Lavoisier in his own right, and form the basis and "fundamental character of the new chemical theory." This announcement, though universally admitted for nearly a century, seems not yet fully comprehended.

Combustion is universally admitted to be a chemical combination of carbon and oxygen, but the heat is supposed to be derived from the carbon. Nearly all writers speak of carbon as producing the heat. Which of these propositions is true ? Does the heat come from the oxygen, or does it come from the carbon? Any body, to give out heat, must be condensed, and all bodies expanded absorb heat. The oxygen, in combining with carbon, is the only body condensed.

Dr. J. R. Mayer, on the sources of heat, says : " It " has been established by numerous experiments that " the combustion of one kilogramme of dry charcoal in " oxygen, so as to form carbonic acid, yields 7,200 units " of heat, which fact may be briefly expressed by say- "ing that charcoal furnishes 7,200 degrees of heat." "Superior coal yields 6,000° ; perfectly dry wood, " 3,300° ; sulphur, 2,700° ; and hydrogen, 34,600° of " heat." In the combustion of hydrogen with oxygen, the heat comes from both gases, the largest amount from the hydrogen. The reason that hydrogen yields the largest amount of heat when condensed, is, that it being the lightest known body, contains the largest amount of imponderable matter, and, when condensed, gives out the largest amount of ether, which, when in excess, becomes heat.

The combustion of oxygen and hydrogen produces more heat than the combustion of any other two bodies, and the reason is, that in these combinations the condensation is the greatest known in chemical combinations, both gases being condensed. The carbon only acts as a condenser of the oxygen ; the heat comes from the oxygen alone in the combustion of charcoal, as Lavoisier truly stated.

The production of heat by condensation holds good in all cases. Matter condensed by mechanical force always gives out heat. This holds true in all forms of matter, gaseous, liquid or solid ; and these bodies, when again expanded, absorb heat. This conception makes clear many modes of producing heat which have been much mystified. For example, in hammering a piece of iron on an anvil, the iron is compressed, and consequently gives out heat ; a bullet is impelled from a gun, and strikes a resisting body, and is condensed, and consequently heated—in passing through the air the air is condensed and gives out heat, and contributes to heat the ball. The smith's hammer compresses the air as well as the iron, and obtains heat from that source.

The production of heat by friction, which destroyed the theory of caloric, and which still bewilders many minds, may be explained by the same theory. The friction caused by rubbing two rough surfaces together not only compresses the solid matter, but likewise the air. If we place oil between the two rubbing plates, so as to exclude the air, the heating is greatly checked ; for this purpose, the axles of the swift traveling cars are always kept well oiled.

The heat and light eliminated by the magneto-electric machine, at first view would appear the most mysterious, and at the same time to give proof of the truth of the dynamical theory. This force appears to be produced by motion alone. No chemical action is here evoked, as in the galvanic pile. If this is not pure motion, and nothing else, what is it? We must bring in the conception long entertained, and now generally coneeded, of the existence of a universal ether; and further, that the universe is a plenum. Even with this conception we are still in the dark. The same motion, in a machine made of wood or stone, would produce neither light nor heat. The motion must be that of a magnet. Here comes up another important question, Why must we have a magnet for this purpose? A magnet is a mass of iron or steel through which a current of electricity or ether circulates. A piece of soft iron, not magnetic, if a current of electricity is forced through it, will answer the same purpose. The flowing current is the necessary condition for the initial action. The force of à thousand horses, to move the machine, would fail to produce the results without the initial current. Here we see that mere motion will not produce the results, and the enquiring mind cannot rest satisfied with this explanation. How, then, are light and heat produced by the magneto-electric machine? Light is now universally admitted to be the result of wave motion of ether, and heat is supposed to result

from some motion in this same ether. The explanation
is found in this theory, that light, heat and electricity
are all manifestations of special motions and conditions
of ethereal matter. Light is the vibratory motion of
this matter; heat is this matter in excess, and electricity this matter in motion. When the machine is
set in motion the current of electricity is established,
and increased in force and rapidity with the increasing
action of the machine. When this current is obstructed,
either by a wire too small to carry it or by charcoal
points, the most intense light and heat are produced.
We accept the dynamical theory, as far as it goes; but
it stops short of an entirely satisfactory explanation.
We accept the theory that heat is a mode of motion,
but we cannot rest satisfied without knowing what kind
of a mode of motion. And more, we want to know
what is moved in the production of heat. We cannot
accept the theory that heat is nothing but motion, for
the reason that we cannot have motion unless something is moved. The addition which we have proposed
to the dynamical theory fills up a blank, and makes the
theory full and intelligible.

The intimate relation which light, heat, electricity
and magnetism bear to each other, are clearly brought
to light. These are all varied effects of one medium.
Interstellar matter is the medium through which all the
forces known as electricity, light and heat are mani-
fested; thus we have one form of imponderable matter,

which receives and imparts all the forces in such varied ways.

Here is the link that binds together all the forces of nature, and makes the universe an inseparable unit. Through this medium the sun enlightens, warms and animates the earth. Through this medium the stars, so far distant as to take thousand of years to transmit their light to us, manifest their powers and glory. Through this medium, we may say, the fingers of God touch the most distant parts of the universe, and connect all creation with the Creator. How beautifully this theory explains and harmonizes the sublime description of God's creation of light, as given in Genesis: "And God said, Let there be light: and there was light." The voice of God set in motion the waves of ether, and these waves, with the rapidity of light, radiated over the whole universe, and there was light thus created. The mind here comprehends the medium through which, and by which, the word created light. Many theologians fear that the advancement of scientific knowledge will lead to skepticism and destroy religious faith. These fears are entirely groundless, for the more profound our knowledge becomes of God's works the higher rises our veneration of that Great, Wise and Good Being who created this wonderful universe in infinite extent and harmony.

3

CHAPTER III.

As we receive all our light and heat either directly or indirectly from the sun, the source of the sun's heat becomes an important problem. This subject has attracted the investigation of the ablest minds, for centuries, without securing a satisfactory solution. We may form a limited conception of the immense amount of heat eliminated by the sun, from a statement made by Lockyer, in his work on astronomy, "that the whole "heat of the sun, collected on a mass of ice as large as "the earth, would be sufficient to melt it in two min- "utes, to boil the water thus produced in two minutes "more, and to turn it all into steam in a quarter of an "hour from the time it was first applied." If the body of the sun contained the amount of heat indicated by his radiation, all the substances known to exist on the earth would be vaporized in a very short time, and the sun itself be converted into a nebulous mass of matter. This vaporization could not be prevented by the greater gravitating force of the sun that binds the particles of matter to his surface.

Tyndall says : "That the heat given out by the sun,
" per hour, is equal to that which would be generated
" by the combustion of a layer of solid coal, ten feet
" thick, entirely surrounding the sun ; hence, the heat
" emitted in a year is equal to that which would be pro-
" duced by the combustion of a layer of coal seventeen
" miles in thickness."

" These are the results, of direct measurement ; and
" should greater accuracy be conferred on them by
" future determinations, it will not deprive them of their
" astounding character. And this expenditure has been
" going on for ages, without our being able, in historic
" times, to detect the loss. When the tolling of a bell
" is heard at a distance, the sound of each stroke soon
" sinks, the sonorous vibrations are quickly wasted, and
" renewed strokes are necessary to maintain the sound.
" But how is its tone sustained ? How is the perennial
" loss of the sun made good ? We are apt to overlook
" the wonderful in the common. Possibly to many of
" us—and even to some of the most enlightened among
" us—the sun appears as a fire, differing from our ter-
" restrial fire only in the magnitude and intensity of the
" combustion. But what is the burning matter which
" can thus maintain itself ? All that we know of cos-
" mical phenomena declares our brotherhood with the
" sun,—affirms that the same constituents enter into the
" composition of his mass as those already known to
" chemistry. But no earthly substance with which we

" are acquainted—no substance which the fall of meteors
" has landed on the earth—would be at all competent
" to maintain the sun's combustion. The chemical
" energy of such substances would be too weak, and
" their dissipation too speedy. Were the sun a solid
" block of coal, and were it allowed a sufficient supply
" of oxygen to enable it to burn at the rate necessary
" to produce the observed emission, it would be utterly
" consumed in five thousand years. On the other hand,
" to imagine it a body originally endowed with a store
" of heat—a hot globe now cooling—necessitates the
" ascription to it of qualities wholly different from those
" possessed by terrestrial matter. If we knew the
" specific heat of the sun, we could calculate its rate of
" cooling. Assuming this to be the same as water—the
" terrestrial substance which possesses the highest spe-
" cific heat—at its present rate of emission the entire
" mass of the sun would cool down 15,000° Fahrenheit
" in five thousand years. In short, if the sun is formed
" of matter like our own, some means must exist of
" restoring to him his wasted power. The facts are so
" extraordinary that the soberest hypothesis regarding
" them must appear wild."

There is another theory, called the meteoric theory.
This theory supposes the sun's heat to be kept up by
meteoric bodies falling on his surface, thus, by dynamic
influence, generating light and heat. Tyndall, in advo-
cating this theory, says : " Were our moon to fall into

" the sun, it would develop an amount of heat sufficient " to cover one or two years' loss ; and were our earth " to fall into the sun, a century's loss would be made " good." If the fall of the earth into the sun would supply its heat for only a hundred years, how long would it take to increase the size of that luminary to such an extent as to change the revolutionary periods of all the planets, and thus derange and eventually destroy the whole solar system ? Tyndall, speaking of the fall of meteors on the earth, says : "At certain " seasons of the year they shower down upon us in great " numbers. In Boston two hundred and forty thousand " of them were observed in nine hours." I saw the same shower of fire. The meteors fell as fast as snow flakes in a heavy storm, and could not be counted. Nearly or quite all of these meteors were consumed before coming in contact with the surface of the earth. No increase of the temperature on the earth's surface was perceived from this extraordinary shower. If we admit the high temperature of the sun claimed by our ablest writers, what would be the effect upon these falling bodies ? Would they not all become vaporized before reaching the sun's surface ? And would not the conversion of solid bodies into vapor carry off or consume a vast amount of heat? We know that water, on being converted into vapor, is expanded eighteen hundred times, and immediately flies off from the surface of the earth. If the surface of the sun is as hot as represented, no foreign body could ever come in contact with it.

Herschel says : " The great mystery, however, is to " conceive how so enormous a conflagration (if such it " be) can be kept up. Every discovery in chemical " science here leaves us completely at a loss, or, rather, " seems to remove farther the prospect of probable ex- " planation. If conjecture might be hazarded, we should " look rather to the known possibility of an indefinite " generation of heat by friction, or to its excitement by " the electric discharge, than to any actual combustion " of ponderable fuel, whether solid or gaseous, for the " origin of solar radiation."

The conception here expressed, that we should look to the electric current as the source of the sun's light and heat, will be found to have received much support from the discovery of magneto-electric currents. The light and heat produced by the magneto-electric machine possess all the properties of the sun's rays. The light and heat thus generated appear to be inexhaustible, and exceed the power of the sun's rays as received on the earth. This current is produced by the revolution of an electro-magnet in the presence of a stationary magnet. May not the sun be so constituted as to be a magnet? In that case, his revolutions on his axis would produce intensely powerful currents. We know that the sun does produce powerful electric currents on the earth, and thus makes the earth an electro-magnet.

Balfour Stewart, in his late work on heat, makes this statement: "It has been discovered by General Sir E. "Sabine that the various disturbances of terrestrial "magnetism are due to the sun, but probably not to his "radiant heat and light. Now, these magnetic disturb- "ances are invariably accompanied by the aurora "borealis, and also by currents of electricity in the sur- "face of the earth, or earth-currents, as they are called. "It would appear, from investigation by the author of "this work, that the earth-currents, and probably the "aurora, are to be regarded as secondary currents, due "to small but rapid changes in the earth's magnetism, "and that the body of the earth may be likened to the "core of a Ruhmkorff's machine, the lower strata of "the atmosphere forming an insulator, while the upper "and rarer, and therefore electrically conducting, strata "may be likened to the secondary coil. In this analogy "the sun may perhaps be likened to the primary cur- "rent which performs the part of producing changes in "the magnetic state of the core."

The magnetic and electric influences of the sun on the earth are being more and more observed, and are attracting the profound attention of our ablest minds. These facts go to show that the sun possesses, in a high degree, magnetic and electric powers. These known and constantly observed electric powers of the sun, to my mind, will more fully explain the nature of the sun's heat than any theory that has yet been advanced.

On the subject of the sun's magnetic influence I quote from Proctor : " The reader will at once see what these " observations tend to. The sun-spots vary in frequency " within a period of ten and a-half years, and the mag- " netic diurnal vibrations vary within a period of the " same duration. It might seem fanciful to associate " the two periodic series of changes together, and, doubt- " less, when the idea first occurred to Sabine it was not " with any great expectation of finding it confirmed " that he examined the evidence bearing on the point. " Judging from known facts, as we may see reasons for " such an expectation in the correspondence of the " needle's diurnal vibration with the sun's apparent " motion, and also in the law which associates the annual " variations of the magnet's power with the sun's dis- " tance. But undoubtedly when the idea occurred to " Sabine it was an exceedingly bold one, and the ridi- " cule with which the first announcement of the sup- " posed law was received, even in scientific circles, " suffices to show how unexpected that relation was " which is now so thoroughly established. For a careful " comparison between the two periods has demonstrated " that they agree most perfectly, not merely in length, " but in maximum for maximum and minimum for mini- " mum. When the sun's spots are most numerous, then " the daily vibration of the magnet is most extensive, " while, when the sun's face is clear of spots, the needle " vibrates over its smallest diurnal arc.

" Then the intensity of the magnetic action has been
" to depend upon solar influences. The vibrations by
" which the needle indicates the progress of those strange
" disturbances of the terrestrial magnetism which are
" known as magnetic storms, have been found not merely
" to be most frequent when the sun's face is most spotted,
" but to occur simultaneously with the appearance of
" signs of disturbance in the solar photosphere. For
" instance, during the autumn of 1859, the eminent
" solar observer, Carrington, noticed the apparition of a
" bright spot on the sun's surface. The light of this
" spot was so intense that he imagined the screen which
" shaded the plate employed to receive the solar image
" had been broken. By a fortunate coincidence, another
" observer, Mr. Hodgson, happened to be watching the
" sun at the same instant, and witnessed the same re-
" markable appearance. Now it was found that the
" self-registering magnetic instruments of the new obser-
" vatory had been sharply disturbed at the instant when
" the bright spot was seen. And afterwards it was
" learned that the phenomena which indicate the pro-
" gress of a magnetic storm had been observed in many
" places. Telegraphic communication was interrupted,
" and at a station in Norway the telegraphic apparatus
" was set on fire ; auroras appeared both in the northern
" and southern hemispheres during the night which
" followed, and the whole frame of the earth seemed to
" thrill responsively to the disturbance which had

" affected the great central luminary of the solar sys-
" tem."

We see, in the above statements, fully illustrated the powerful magnetic influence of the sun, and a strong confirmation of the conception that the sun is a magnetic body, and, as such, has the power to generate intense heat by his own motions. One remarkable effect was shown, in the case just stated, that the magnetic current produced intense heat, sufficient to set fire to the telegraphic apparatus. This impulse which came from the sun to the earth possessed all the properties of magnetism, electricity and heat. As the sun possesses magnetic and electric forces within himself, and as these forces, when obstructed, produce the most intense heat known, he must possess the power to produce his own supply of heat, without limit or exhaustion. These forces are manifest on the earth in consequence of the actions on the sun's surface.

If no matter existed between the earth and the sun his impulses could not be felt on the earth. These impulses travel with the speed of light, and reach us in a very brief space of time. This shows that the atoms of this matter must be very close to each other, or in actual contact, and they must be solid and unyielding in form, or else these impulses could not be transmitted so rapidly. All late and able observers agree in stating that the sun's atmosphere is in an intense state of action, and all know that the influence of these actions is intensely felt on the earth's surface, and even through the body of the earth. The simple revolution of the sun on his axis, unless possessing electro-magnetic powers, would not produce these results, but the sun acting as a magnet these results would follow as they do in our magneto-electric machines. On the earth we have to furnish the mechanical power to move the machine ; the sun has his own permanent motions, and consequently no material is consumed. How this motion was first given to the sun is, and ever will be, only known to God. We know that the sun has his motions, and we have strong reasons to believe that these motions will be indestructible. This is all that we require to give consistency to our theory. The magnets, if connected with good conductors, do not become heated by their motions, though sending off the most intense light and heat known on the earth. May not the sun, in the same way, remain cool whilst sending off these intense streams of light and heat.

The elder Herschel believed the body of the sun to be cool, and a fitting place for the maintenance of life. The indications given by the sun spots are that the body of the sun is darker than his surrounding photosphere. This appearance would indicate that the body of the sun may be cooler than his photosphere.

When we examine closely the condition of the planetary bodies that revolve around the sun, we find they have each some mode by which their temperature is regulated to fit them for the maintenance of life.

The greatly varying distances of the planets from the sun would seem to indicate that the temperature of each greatly differs. The earth, for example, at the distance of 91,430,000 miles, has a temperature just suited for the maintenance of animal life. Mercury, at the distance of only 35,393,000 miles, might be supposed to be intensely heated by the sun's rays, and thus unfitted for the maintenance of life in any form. As this planet is so nigh to the sun, astronomers have been unable to make observations that either prove or disprove this conjecture. Venus, at the distance of 66,131,000 miles, would still be too hot for the maintenance of life. Analogy, however, leads us to believe that this may be a habitable globe. Mars, at the distance of 139,312,000 miles from the sun, should be, judging from his distance from the sun, a frozen mass, too cold for life to exist on his surface. There fortunately astronomers have been able to determine his condition.

The climate of Mars must be about the same as that of the earth, notwithstanding the light and heat he receives is less than one-half of that received by the earth. Observations have shown that he is surrounded with vapors of water—that he has clouds, with rain and snow. The telescope plainly shows that his poles are covered with snow and ice, and that the boundaries of these snows are constantly changing as the sun approaches to or recedes from them, in the same way as on our earth. Some mode has been provided by the Creator by which this planet has been made a fitting habitation for man and other animals. If the earth received only the same amount of heat and light that Mars does, it would be a dead and frozen mass. The bearing which these, and other, more remarkable facts, may have in explaining the nature of heat, will be more fully seen hereafter.

Jupiter is distant from the sun 475,693,000 miles. The light and heat which he receives from the sun are reduced to about one twenty-fifth of our supply on the earth. This supply of heat is so small that Jupiter, according to calculation, would be nothing but a frozen iceberg.

Saturn's distance from the sun is 872,135,000 miles. This light and heat are reduced to one ninety-first part of that received by the earth. And yet the telescopic appearances of these planets indicate that their temperature cannot vary much from that of the earth. Belts of clouds are always seen by the telescope to encircle these two orbs, and these clouds are constantly changing form. Here water remains unfrozen, and much of it in a state of vapor; consequently the climate must be nearly the same as ours.

Uranus is distant from the sun 1,752,851,000 miles; and Neptune 2,746,271,000 miles distant from the sun. The amount of light and heat which Uranus receives is one three hundred and ninetieth, and that of Neptune one nine hundredth, of that of the earth. These planets are so very far distant that the telescope gives us no indication of clouds or vapor.

The theory of heat which we are here advancing, has an important bearing on the question. How are the temperatures of these planets kept to resemble each other so closely?

In the first place, we can imagine the universe as full of matter; all the planetary bodies moving in a universal ocean of imponderable matter, this matter possessing no force of gravity, free to move in every direction alike; and moved in various directions by the impulse of all moving ponderable matter. We may suppose that heat is this imponderable matter in excess, that is caused by the vibratory motion of this ether; that electricity is this matter in motion, and that magnetism results from the flowing current of electricity.

We can comprehend how Mars, though receiving less than half the light and heat which the earth receives may be modified by the vapors that surround him, so as to make his temperature resemble closely that of the earth.

The vast distance of Jupiter from the sun forbids the supposition that the light and heat which he possesses is all received from the sun. He lies more than five times farther from the sun than our earth, and the light and heat which he receives from that orb are reduced to about one twenty-fifth of our supply. This amount of heat is entirely too small to produce the actions that are known to be going on on his surface. The enormous masses of vapors that are floating in his atmosphere indicate a temperature quite equal to that of our earth.

Proctor in writing on this subject says : " It seems to " me, that these considerations point with tolerable " clearness to the conclusion that, within the orb which " presents so glorious an aspect upon our skies, processes " of disturbance must be at work wholly different from " any taking place on our earth."

" That enormous atmospheric envelope is loaded with " vaporous masses by some influence exerted from be- " neath its level." " Those disturbances which take " place so frequently and so rapidly are the evidences of " the action of forces enormously exceeding those which " the sun can by any possibility exert upon so distant a " globe." " When we see masses so enormous, swayed " by influences of such energy, that intermediate belts, " thousands of miles in width, are closed up in a single " hour; when we recognize the tremendous character " of the motions which from beyond four hundred mil- " lions of miles, are distinctly cognizable by our tele- " scopes, we see that we have no ordinary phenomena " to deal with, and that the theory we adopt for the " explanation cannot be otherwise than striking and " surprising."

The late Professor G. Bond, calculated that Jupiter

sends forth more light than he receives from the sun. Jupiter must have some power within himself by which a large portion of his light and heat is generated. The light sent to us from Saturn also bears a much greater proportion to the amount of solar light actually received by the planet than is observed in the case of Mars or the moon, and so nearly approaches the proportion noticed in Jupiter as to lead to the same inference— namely, that a portion of Saturn's light is emitted from the body of the planet. If we admit that Jupiter and Saturn generate a portion of their light and heat, they in this case resemble the sun. Their densities are nearly the same as the sun's. If the sun is so constituted as to resemble a magnet, so may be the superior planets.

The rapid revolutions of the superior planets on their axis, which is performed in less than half the time of that of the earth, would give them greatly increased electric powers. The temperatures on the planets can not differ very widely, though varying so extremely in distances ; as we recognize the vapors of water in all except the two most remote, Uranus and Neptune. How can these nearly equal temperatures be maintained under these widely differing conditions ? We can readily imagine that the conditions of the earth and Mars may be so regulated by the vapors surrounding them as to make their temperatures nearly the same, but the superior planets are so far removed from the

Sun as to make it nearly impossible for them to receive
their light and heat all from that distant luminary.

Proctor, and many other eminent astronomers, believe
that the superior planets send out more light and heat
than they receive from the sun; and that they are hot
bodies cooling down. We have already alluded to this
theory in regard to the sun, and have found it liable to
very grave objections.

The theory that these bodies, like the sun, are so
constituted as to possess magnetic powers, will give a
conception of the mode by which all these forces may
be regulated. To say that all these forces are regulated
by electric currents will not satisfy the enquiring
mind, for the question comes up unbidden, what is
electricity?

If we say that electricity is a flowing current of imponderable matter, that fills all space, we have a conception on which the mind can rest; and which, I believe,
will harmonize and explain all the known forces of
nature. If the existence of this matter is admitted,
what causes the motions that are taking place therein?
That all the masses of ponderable matter in the universe
are in regular and constant motion we all know. The
origin of this first impulse can only be known to God.
We take the known motions of the solar and stellar
universe, and trace from them the cause of all motions in
universal matter. Heat, light, electricity and magnetism,
all result from the motions of the solar and stellar

4

worlds through this medium, universal imponderable matter.

As we know that nearly all the forces operating on the earth are derived from the sun, the question of the indestructibility of force must be determined by the properties of the sun's force. Are the sun's forces permanent and unchangeable?

The nebular theory of the formation of the universe, adopts the conception that all the stellar and solar worlds are, and have been for countless ages, cooling down; and that the sun and planets are still undergoing this process. If this conception be true, the sun's as well as all other forces must be destructible.

The geological condition of the earth bears evidence, that all its parts have been, in different periods of time, subjected to intense heat. This fact is cited in proof of the nebular theory. We must not forget that the condition of the earth also shows, that all its parts, have been, in different periods of time, subject to much more intense cold than that which now prevails. Either the earth must have been heated and cooled in parts, or there must be some compensating forces that restored that which had been lost.

That the forces of the universe are compensating, we have evidence from some of our ablest astronomers and mathematicians. Sir John Herschel says: "The move-"ments of the perihelia, and variations of eccentricity of "the planetary orbits, are interlaced and complicated

" together in the same manner and nearly by the same
" laws as the variations of their nodes and inclinations.
" Each acts upon every other, and every such mutual
" action generates its own peculiar period of compensa-
" tion, and every such period is thence propagated
" throughout the system. Thus arise cycles upon
" cycles, of whose compound duration some notion may
" be formed, when we consider what is the length of one
" such period in the case of the two principal planets—
" Jupiter and Saturn. The greatest eccentricity of
" Jupiter corresponding to the least of Saturn, and
" vice versa. The period in which these changes are
" gone through would be 70,414 years. After this
" example, it will be easily conceived that many millions
" of years will require to elapse before a complete fulfil-
" ment of the joint cycle which shall restore the whole
" system to its original state as far as the eccentricities
" of its orbits are concerned."

" Now, it may naturally be inquired whether, in the
" vast cycle above spoken of, in which, at some period
" or other, conspiring changes may accumulate on the
" orbit of one planet from several quarters, it may not
" happen that the eccentricity of any one planet—as
" the earth—may become exorbitantly great, so as to
" subvert those relations which render it habitable to
" man, or to give rise to great changes, at least, in
" the physical comfort of his state. To this the
" researches of geometers have enabled us to answer in

" the negative. A relation has been demonstrated by
" Lagrange between the masses, axes of the orbits, and
" eccentricities of each planet, similar to what we have
" already stated with respect to their inclinations, viz :
" that if the mass of each planet be multiplied by the
" square root of the axis of its orbit, and the product by
" the square of its eccentricity, the sum of all such pro-
" ducts throughout the system is invariable ; and as in
" point of fact, this sum is extremely small, so it will
" always remain. Now, since the axes of the orbits are
" liable to no secular changes, this is equivalent to saying
" that no orbit shall increase its eccentricity, unless at
" the expense of the common fund, the whole amount
" of which is, and must forever remain extremely
" minute."

We may reasonably conclude that the sun has within
itself compensating forces that generate its light, heat
and magnetic force ; and that this force will be perpetual.
Cycles of variations are constantly generated, but these
as above stated have their compensations which keep
their powers unwasted.

After the great discovery of the indestructibility of
matter, philosophers conceived that force must also be
indestructible. This conception cannot be sustained
when we apply it to the forces acting on the earth, for
we know that the earth's forces are constantly wasted,
as in the radiation of heat, and in many other ways.
The earth's forces are nearly all derived from the sun,

and are renewed by the sun, as fast as wasted. The sun's forces being indestructible, we still have, fully established, the two grand conceptions—the indestructibility of matter and the indestructibility of force. Then follows, as a necessary consequence, the everlasting duration of the *universe.*

Balfour Stewart in his late work on "The Con-"servation of Energy," uses the following language : 'Although, therefore, in a strictly mechanical sense, " there is a conservation of energy, yet, as regards use-" fulness or fitness for living beings, the energy of the " universe is in process of deterioration. Universally " diffused heat forms what we may call the great waste-" heap of the universe, and this is growing larger year " by year. At present it does not sensibly obtrude " itself, but who knows that the time may not arrive " when we shall be practically conscious of its growing " bigness?". This conclusion is based on the supposition that there is such a thing as a hot fluid, and that it is passing from the earth and sun to enter, and remain in, universal space; in other words, that the sun and earth are losing their heat, and cooling down. We have no evidence to show that the sun or earth is any colder now than it was thousands of years ago ; on the contrary the facts show that their heat has ever been a fixed quantity.

Many able writers represent the earth as much hotter during the carboniferous than the present period ; but recent geological discoveries make this hypothesis untenable. Impressions of rain drops have been detected in carboniferous sandstone by Dr. Dawson, Sir Charles Lyell, and more recently by Mr. Brown, in Australia ; and these rain-marks are, on the average, about as large as those which are produced by the rain of our own period. As Lyell well remarks, " The great humidity " of the climate of the coal period had been previously " inferred from the number of ferns, and the continuity " of its forests for hundreds of miles ; but it is satis- " factory to have at length obtained such positive proofs " of showers of rain, the drops of which resemble in " their average size those which now fall from the " clouds. From such data, we may presume that the " atmosphere of the carboniferous period corresponded " in density with that now investing the globe, and that " different currents of air varied then as now in tem- " perature, so as to give rise by their mixture, to the " condensation of aqueous vapour." These marks of rain-drops on the solid rocks, small and simple as they may appear, give evidence to my mind that the heat of the earth during the carboniferous period must have been nearly the same as at the present time. I believe that nearly all geologists agree that millions of years have rolled around since the carboniterous strata were formed. When we state that the earth's heat has not varied during our historic period, we are met by the statement that six thousand years is too short a period

to test this question. Here we have a period of millions
of years testifying to the same great truth : that the
earth's heat has been a fixed quantity for millions of
years.

The conception which I have advanced regards heat
as the universal ether, and the condition which constitutes heat as this ethereal matter in excess.

This ether at rest is not heat. When heat is radiated
into space, it is not lost nor wasted, but remains a part
of the great ethereal ocean. This ocean of matter is
being constantly disturbed by the motions of all ponderable matter, as in the revolutions of all the heavenly
bodies. These disturbances produce currents in this
great ocean. When these currents are obstructed, as by
coming in contact with the earth, excess of ether, or
heat, is the result. The earth in turn radiates its
heat into space ; but as the sun constantly pours these
streams on the earth, the earth's heat is kept a constant quantity.

These streams if not obstructed are not hot, as we
find on the tops of high mountains, or in the upper
regions of the air. On the tops of high mountains
where the air is thin, and does not obstruct the sun's
rays the air remains very cold ; but when these rays
light on a person in these situations, the heat is very
intense. The rays of the sun, unless absorbed, are not
hot. If they should pass through a perfectly transparent
body, or be perfectly reflected from a polished surface,

no heating effect would take place in these bodies. All these facts show that heat is ether in excess.

If this ether could attain to a perfect state of equilibrium, such forces as heat, light and electricity could not exist. Here then, again comes in the question: can, or will this state of equilibrium ever take place? We certainly never can have this state of equilibrium as long as there are bodies of ponderable matter moving in this great ocean of ethereal matter.

As the waters of the ocean are raised from their bed and sent in streams of vapor through the air, and return again in streams to the ocean, so the ether is sent in streams in various ways, and returns again to the great ocean whence it came. No one supposes that one drop of the waters of the ocean is lost. Then why should we suppose that these streams of ether are lost, or become waste-heaps in the universe? The glacial evidences as seen on the rocks, all over the earth, seem to indicate that the earth was once colder than it now is. Cycles of changes appear to have occurred on the earth, but compensating forces have restored the temperature to what we may call a normal condition.

Writers cite the moon as an example of a body that has lost all of its original heat. The moon, as is well known, has its sides at intervals, intensely hot and intensely cold, as its sides are alternately turned to or from the sun. The moon's heat, as well as that of the earth, is a fixed quantity.

The conception that heat is an igneous fluid, leads us to the conclusion that the energies of the universe are wasting. The conception that heat is imponderable matter in excess, caused by the motions of ponderable bodies moving in the ocean of universal ether, allows the universe to be indestructible.

The conception that imponderable matter would retard the motions of ponderable matter moving through it, would make the universe subject to decay.

Would matter devoid of gravity resist the motions of ponderable matter? That matter which has no gravitating force would be free to move in any direction, and consequently would not resist the impulses of matter endowed with gravitating force. If the heavenly bodies move in a resisting medium their motions must, certainly, be extinguished at some time. If they move in a medium without resistance, their motions must be perpetual : so we see that the question of the conservation of the forces of the universe, rests on this single point. I believe that matter without the force of gravity, will not resist the motions of ponderable matter; and on this conviction base my faith in the stability of the universe.

CHAPTER IV.

The very large amount of coal that is now being used for the various purposes of life is beginning to excite the attention of many, and the anxiety of some persons. The British Parliament in 1870, appointed a commission to enquire how long their coal fields would last at the present rate of consumption. That commission reported that the last year they raised from the mines of Great Britain one hundred and ten million tons. The vastness of this quantity may be shown by stating that the diameter of the earth is 7,926 miles, or 13,880,760 yards; the coal raised in 1870 would make a solid bar more than eight yards wide and one yard thick, which would pass from east to west through the earth at the equator.

It was reported to the House of Commons by a member of the Coal Commission, that the decision of that body, after long and laborious inquiry, would be that there existed in their coal fields, a supply for about one thousand years at our present rate of consumption. This rate, we must remember, is rapidly increasing. This increasing rate of consumption will bring down the period of duration to less than five hundred years. The British people are beginning to complain of the increasing price in consequence of the increasing difficulty of deeper mining.

This question of the supply of heat and light for the inhabitants of the earth, is one of the deepest importance. What will be the effect on the inhabitants of the earth when the coal fields are used up? Would the human race descend again into barbarism? Would the bronze age, and the stone age return, and men again retire to the caves in the earth for warmth? Without coal or wood, we could no longer extract the iron from the ore, or melt that already extracted to give it new shapes to meet our varied wants. We could no longer drive our railroad cars over the earth, or our steamships over the ocean. The steam mills in their varied forms that now do the work of millions of men and horses, would become as silent as death, and the beautiful and useful products in their myriad forms, that now give so much happiness to the race, would no longer be found.

Shall we look into the future and anticipate these gloomy prospects as likely to fall on our descendants, or shall we rather confide in the wisdom and goodness of the Creator who framed this wonderful and beautiful world? That being who created this world, and the whole universe, with so much wisdom and skill, and stored it with such rich, wonderful and varied blessings to add to the happiness of its inhabitants, surely does not intend to leave it to desolation and ruin. Already the light of science is beginning to give us light and hope. The advancing footsteps of scientific knowledge are beginning to reveal to our vision, prospects in the future far more glorious than the past. The discoveries that have been made on this subject that we have thus far been attempting to discuss, light and heat, show us that we have sources of light and heat that far excel all that has been produced by the use of coal.

The galvanic battery can be made far to excel the best wind furnace in the production of heat, and will fuse platinum like wax. Quartz, sapphire, magnesia and lime are all fused by this heat. Carbon is the only substance which cannot be melted by the pile, though with six hundred Bunsen cells it has been softened to such a degree that adjoining pieces will adhere; which seems to indicate the commencement of fusion. The metals are not only melted but volatilized, and dissipated in vapor.

The light produced by the galvanic current is equally remarkable with that of the heat; it can be made to rival that of the sun.

The magneto-electric current gives more promise than the galvanic battery. Machines have been constructed which give both light and heat of the most intense power. From experiments performed in the month of June, with Wild's magneto-electric machine, comparisons of sun light with the electric light armed with the reflector, by means of the shadows thrown by both from the same object, the electric light seemed to possess three or four times the power of sun light. That the relative intensity was somewhat in this proportion, was evident from the powerful scorching action of the electric light on the face, and the ease with which paper could be set on fire with a burning glass when introduced into its rays. Light enough could be produced by this machine and lamp, if placed on the top of a high tower, to illuminate London by night, more brightly than the sunlight does by day.

One great advantage of this machine is its capability of enlargement to any required power. If instead of using the current from the ten-inch armature of the second electro-magnet for the production of light, it were to be used in producing a still larger electro-magnet, a vastly greater development of power would be the result. The only apparent limit to this multiplication of power, is the excessive heat which would be developed in the rotating armatures; this might, perhaps, be pushed so far as to burn up all the working parts, dissipate the electric lamp and conducting wires, destroy the attendants, and become in fact perfectly unmanageable.

Vast improvements will be made in the construction of machinery for utilizing these forces, and making them subservient to man's wants. The time will probably come, when man will obtain all the light and heat which he may require for the practical purposes of life, from chemical and electrical sources.

The time may come when our descendants will look back on our condition as we look back on the condition of our ancestors who inhabited the earth during the stone age, and wonder how we ever managed to live with all the discomforts of stoves, coal ashes, and coal gas filling our dwellings, and all the labor of building fires and cleaning dust. Only think how convenient it would be to have our houses heated and lighted by an electric current; no smoke, no dust, no offensive gas; and then our cooking all could be properly done without cooking the cook. Think of our cities with one great central lamp and all our streets as light at midnight as at noon day; in this case no lights would be required in our houses to light our rooms. The exhaustion of our coal fields may, after all our fretting, be a great blessing to the human race.

That being who has formed and adjusted the universe with so much wisdom and goodness, surely does not intend that his works shall come to nought. I cannot believe that the material universe is wearing out; or that the forces of nature will ever become exhausted. Some eminent writers speak of the debris and waste-heap of the universe. The flying atom, as well as the rolling world, is fulfilling its ordained mission. We see the earth's forces constantly changing, and constantly becoming exhausted, but we see them constantly renewed by the forces of the sun.

The earth set off by itself, would soon become a dead and frozen mass of matter ; but the earth is not thus set off; it is a constituent atom of the universe, and as such is subject to the universal and everlasting laws. Compared with the whole, it is but an atom, yet its annihilation, if that were possible, would probably derange the universal harmony.

"From Nature's chain whatever link you strike,
"Tenth or ten thousandth, breaks the chain alike.
"And if each system in gradation roll,
"Alike essential to th' amazing whole,
"The least confusion but in one, not all
"That system only, but the whole must fall.
"Let earth, unbalanced from her orbit fly,
"Planets and suns run lawless through the sky;
"Let ruling angels from their spheres be hurled,
"Being on being wrecked, and world on world;
"Heaven's whole foundations to their centre nod,
"And Nature trembles to the throne of God."

CHAPTER V.

INFLUENCE OF THE VITAL FORCES.

The production of heat by the vital forces is more difficult to comprehend than by the ordinary physical laws. Much vital heat is, no doubt, produced by a purely chemical process; as in breathing—the oxygen of the air combining with the carbon of the blood, under the same laws as in ordinary combustion. This purely chemical process will account for much of the vital heat, but by no means will it account for the large and constant supply of heat found in vital organisms. The amount of carbon and oxygen consumed would be entirely too small to keep a living animal steadily at a temperature of ninety-eight degrees Fahrenheit The human body has a system of nerves which exert a powerful influence in the production and distribution of the heat of the body. Muscular contraction, also, exerts a great influence in the production of heat.

In what way does muscular contraction produce heat? The muscles are, certainly, not disintegrated by their actions, and consequently these actions cannot be purely chemical. These actions cannot be accounted for by the supposition that the muscles occupy less space when contracted; for what is lost in length is gained in thick-

ness. How, then, can we account for the production of
heat by the contraction of the muscles?

The human body has a system of nerves which control the distribution of heat to all parts of the body, and
keep up an equable temperature over the whole body.
These nerves are found to be perfect conductors of electricity. A current of electricity will pass rapidly
through these nerves, and exert many of the influences
of the nervous fluid.

The nervous system is a living magneto-electric
machine, the nerves acting as a magnet, the muscles
giving the motions required to produce the effects. The
conducting nerves carry the excess of heat from the
interior of the body to the surface, where it can be car-
ried off by radiation or by the evaporation of the fluids.
The power which the living system possesses to regu-
late its own temperature, cannot be accounted for on
strictly chemical or dynamical principles. We can
readily account for the production of heat, but the regu-
lation of that temperature, under such widely differing
conditions, is profoundly mysterious. The temperature
of the human body remains nearly the same when ex-
posed to heat of one hundred degrees Fahrenheit, or to
cold below zero. This wonderful power cannot be
accounted for by any of the laws governing inorganic
matter.

Even in the vegetable world the vital force plays an
important part in preserving the organism. During the

5

last winter the delicate peach trees in my yard were subject, for many days, to a temperature below zero ; and yet they are now budding forth in full vitality.

This instinctive intelligence of the vital forces cannot be accounted for under any of the laws governing inorganic matter. The human body maintains a temperature of nearly ninety-eight degrees Fahrenheit in all latitudes of the earth. It possesses the power to lessen excesses, and to supply deficiencies. How can this remarkable power be accounted for? No hypothesis has yet been advanced that will in any way explain these mysterious actions.

The conception which has been advanced in this thesis will give us a probable explanation. When we admit the existence of imponderable matter filling all space, and the spaces in all matter, we have taken the first step towards an explanation. When we admit that this matter in excess constitutes heat, we have taken the second step ; and when we admit that motion will disturb the equilibrium of this matter, we have arrived at a probable solution of this profound problem.

Any motion of matter in the great ocean of matter must disturb the equilibrium, and consequently produce currents. These currents, when obstructed, will cause excess of this matter at points where obstructed, consequently heat will be produced. In saying that heat will be produced, I do not mean to say that we have produced caloric, or an igneous fluid, only we have produced that condition of imponderable matter which we have previously stated constitutes heat. The tree that preserves its life at a temperature below zero does not do it by chemical actions, as no such actions are going on at that temperature. Under these circumstances there is neither growth nor decay going on, consequently heat cannot be produced in this way.

We know that the sun produces electric currents in the earth, and that these currents prevail in all seasons of the year. We know that electric currents obstructed produce heat. Matter in some conditions conducts electric currents, and in other conditions obstructs electric currents. A large copper wire will freely conduct a large current of electricity, but if this wire should connect with a non-conducting substance intense heat would be produced. The tree in the ground may, and probably does, by its peculiar vital organisms utilise the earth currents of electricity to preserve its vitality under various temperatures.

The preservation of animal temperatures may be accounted for in the same way. The nerves and muscles are good conductors, but the intervention of non-conductors at any particular point would produce heat. The nutriment of the body that flows from the stomach to the lungs, and from the lungs to all parts of the body and is deposited just where it is required, illustrates the flow of heat, which is distributed by the same influence, and with the same intelligence and wisdom.

Bone is deposited where bone is required, and muscle where muscle is required, and nerve where nerve is required, and so on in the whole process of nutrition. Heat is produced and distributed under the same general intelligent laws.

The forces governing organic matter are directly the opposite of those governing inorganic matter. Life is a constant struggle between these two opposing forces; as long as the vital forces prevail life is preserved. When the inorganic forces prevail death ensues, and the body that was animated with life is decomposed. The great vital machine no longer produces and regulates its own heat, but is subject to all the changes of temperature that surround it.

We cannot come to the conclusion, with all these facts before us, that the same laws govern living matter that govern inorganic matter. The actions producing heat in the living body cannot properly be called combustion. Nor can we properly say that digestion and assimilation are purely chemical processes. The vital forces control one set of actions, and the chemical forces the other.

When we contemplate the actions of the vital forces we find ourselves surrounded by impenetrable mystery. Here science fails to guide us into the open fields of knowledge. We perceive the mysterious actions that are going on around and within us, but we cannot perceive the secret springs that animate these actions. We can understand the mechanical construction, and design of the living machine—its purposes and objects, but when we seek for the cause of the motive power we find ourselves completely in the dark.

The vital forces appear to act with intelligence; they tend to accomplish designs and purposes. The human body is built up with more intelligent design than any machine that man's intelligence has ever constructed. The adaptation of means to ends is most perfect. The vital forces not only build up these wonderful structures but they constantly tend to repair and preserve them from decay. Each organism has its specific period of duration; up to that time the vital forces all tend to preserve and repair the organisms. The vital forces add intelligence to matter and motion. This intelligence is not material, but becomes a controlling force over matter.

What is this force which takes control of matter and makes it conform to intelligent designs and purposes? This question leads us above and beyond scientific knowledge. The mind of man feels and knows that it has powers of thought and reason, and that this thought is not material. Thought is not bounded by time or space—it traverses the whole universe in a moment. Light, the most swiftly traveling agent known, will take thousands of years to pass from some of the most distant stars to us; and yet thought will in a moment pass this almost infinite space.

Many theologians fear that the study of the natural sciences will lead to materialism. A superficial study of these subjects may lead to these results, as we see matter and motion controlling so much of the universe; but on a full investigation of material things we are compelled to supplement these with something that is not material.

Intelligent design, order and harmony are everywhere seen throughout the universe. The creation was ordained with intelligence and wisdom, and is controled and sustained by intellectual power. The universe is more than matter, and force. Intelligence rises over and above all material things, and controls and sustains all with intelligent designs and purposes. This intelligent design is not only seen in the control of unnumbered worlds, but we can trace it down to the smallest animalcule, and smallest atom of matter. *The first and highest power in the universe stands intelligence.*

Man is not the mere creature of accident, but is created by an all-wise good and powerful creator, with intelligent design and purpose. Here we may rest our faith, here we may base our trust, and work on with full confidence in the wisdom and goodness of the great creator of the universe.

Intelligence is not only displayed in the mechanical and useful, but in the beautiful; nature everywhere strives to supply the absolute wants of man and to minister to his pleasures. Who that looks over the earth, clothed with life and beauty, can fail to see the wonderful working of this mysterious power. Vegetable life decorated with more than royal splendor, and with more variegated and beautiful colors than the rainbow— insect and animal life vie with the vegetable in ministering to our happiness through our sense of beauty. Nature everywhere more than realizes the beautiful conceptions of the poet :

> "Warms in the sun, refreshes in the breeze,
> "Glows in the stars, and blossoms in the trees;
> "Lives through all life, extends through all extent,
> "Spreads undivided, operates unspent."

9 7 8 3 7 4 3 4 1 7 4 3 4